by Adrian Harrison

INTRODUCTION TO

POLYNOMIALS

January 2020

CONTENTS

POLYNOMIALS

Definition

A polynomial is any function f of the form

$$P(x) = a_0.a_1x.a_2x^2 ....a_nx^n \text{ where}$$

$a_0.a_1.a_2............a_n$ are real numbers ($\in R$) and where n

Is a natural number.

Example:

1. $P(x) = 4x^6 - 3x^{-2} + 6x_{+1}$

$P(x)$ is not a polynomial because the power of

$-3x^{-2}$ is -2,but -2 is not a natural number.

2. $P(x) = 6x^5 + 7x^3 + x^{\frac{1}{2}} + 3$

$P(x)$ is not a polynomial because of the power "$\frac{1}{2}$",

"$\frac{1}{2}$" is not a natural number.

Example:

($P(x)$ is a polynomial)

$$P(x) = ax^6_{+(a-b+3)}x^{-3} + (a + b - 9)x^{-2} + bx$$

$\Rightarrow P(x) = ?$

A)$2x^6 + 3X$ 　　　　　　　　　　B) $3x^6 + 6X$ 　　　　　　　　　　C) $5x^6 + 4X$

D) $2x^6 + 5X$ 　　　　　　　　　　E) $3x^6 + 9X$

(Solution):

$$a - b + 3 = 0$$
$$\underline{+\ a + b - 9 = 0}$$
$$2a - 6 = 0$$

a=3,(and)b=6

$\Rightarrow P(x) = 3x^2 + 6$

-Answer B

(Example)

(P(x) is a polynomial)

$P(x^2)=(a-2)x^5 + ax^4 + (b-4)x^3 + 2bx^2+3b$

$\Rightarrow P(x) = ?$

A)$2x^2$+8X+12 　　　　　　　　B) $2x^2$+4X+1 　　　　　　　　C)
$4x^2$+12X+6

D) $3x^2$+5X+8 　　　　　　　　　E) $2x^2$+6X+4

(Solution):

$P((\sqrt{x})^2) = (a\text{-}2)$
$(\sqrt{x})^5 + a(\sqrt{x})^4 + (b-4)(\sqrt{x})^3 + 2b(\sqrt{x})^2 + 3b$

$$\Rightarrow P(x) = (a-2)x^{\frac{5}{2}} + ax^2 + (b\text{-}4)x^{\frac{3}{2}} + 2bx + 3b$$

$a\text{=}2\text{=}0 \Rightarrow a = 2$

$b\text{-}4\text{=}o \Rightarrow b = 4$

$P(x) = 2x^2 + 2.4x + 3.4$

$= 2x^2 + 8x + 12$

-Answer

A

(Example):

$$x^3.P(x) = ax^8 + (b-2)x^5 + (a-4)x + b - 5 \Rightarrow P(x) = ?$$

A) $3x^5 + 2x^2$ B) $3x^2 + 4x^2$ C)
    $4x^5 + 3x^2$

D) $8x^5 + 2x^2$ E) $7x^5 + 2x^2$

(Solution):

$$\frac{x^3 P(x)}{x^3} = \frac{ax^6}{x^3} + \frac{(b-2)x^5}{x^3} + \frac{(a-4)x}{x^3} + \frac{b-5}{x^3}$$

$P(x) = ax^5 + (b-2)x^2 + (a\text{-}4)x^{-2} + (b-5)x^{-3}$

a-4=0 (and) b-5=0

a=4, b=5

$P(x) = 4x^5 + (5-2)x^2$

$= 4x^5 + 3x^2$

-Answer C

(Example):

$P(x)=-3x^3 + 4x^2 - X + m + 1$

$P(2)=4 \Rightarrow m = ?$

(Solution):

$P(2)=-3.2^3 + 4.2^2 - 2 + m + 1 = 4$

-3.8+4.4-2+m+1=4

-24+16-2+m+1=4

-9+m=4

m=9+4

m=13

(Example):

$P(x)=5x^6 - 4x^3 + 11 \Rightarrow$ $\qquad$ $P(\sqrt[3]{2}) = ?$

(Solution):

$P(\sqrt[3]{2}) =5.(\sqrt[3]{2})^6 - 4(\sqrt[3]{2})^2 + 11$

=5.4-4.2+11

=20-8+11

=23

(Example):

$n \in Z$

$P(x)=5.(x - 2)^{2n} - 7(2 - x)^{2n - 1} \Rightarrow P(1) = ?$

(Solution):

$P(1)=5.(1 - 2)^{2n} - 7(2 - 1)^{2n - 1}$

$\quad = 5.(- 1)^{2n} - 71^{2n - 1}$

$\quad = 5.1 - 7.1$

$\quad = 5 - 7$

$\quad = -2$

EQUALITY OF POLYNOMIALS

$P(X) = a_n x^n + a_{n - 1} x^{0 - 1} + \ldots + a_1 x + a_0$

$Q(x) = b_n x^n + b_{n - 1} x^{n - 1} + \ldots + b_1 x + b_0$

$P(x) = Q(X) \Rightarrow a^n = b_n, a_{n - 1}, \ldots a_1 = b_1, a_0 = b_0$

(Example):

$P(x) = (x^2 + 1).(x+3)$

$Q(x) = x^3 + ax^2 + bx + c$

$P(x) = Q(x) \Rightarrow a = ?, b = ?, c = ?$

(Solution):

$P(x) = (x^2 + 1)(x + 3) = x^3 + ax^2 + bx + c = Q(x)$

$P(x) = x^3 + 3x^2 + 1x + 3 = |x^3 + ax^2 + bx + c = Q(x)$

$\Rightarrow a = 3, b = 1, c = 3$

5

(Example):

$P(x)=(x-2)(x^2 + px + 3) + x - 5$

$Q(x)=x^3 + 3x^2 + bx + c$

$P(x)=Q(x) \Rightarrow b + p = ?$

A)-5   B)-4   C)-3

D)-2   E)-1

(Solution):

$P(x)=x^3 + x^2(p - 2) + (4 - 2p)x - 11$

$Q(x)=x^3 + 3x^2 + bx + c$

$P(x)=Q(x)$
$\Rightarrow p - 2 = 3 \Rightarrow p = 5, 4 - 2p = b \Rightarrow b = 4 - 10 =- 6, c =- 11$

b+p=5-6=-1

SUM OF COFFICIENTS ON POLYNOMIALS

A polynomial P(x) is given

1.To find the sum of the coefficients of P(x).write 1

Instead of x.

2.To find the constant term of P(x).write 0 instead of x.

P(0)=Constant term

(Example):

$P(x)=(x^2 - x + 2)^3(x^4 - 2x + 4)^2$

What is the sum of the coefficients of P(x)?

A)48   B)66   C)72   D)84   E)90

(Solution):

X=1

$P(1)=(1 - 1 + 2)^3.(1 - 2 + 4)^2$

$=(2)^3.(3)^2 = 8.9 = 72$

-Answer C

(Example):

$P(x^3 + 8) = x^6 - 2x^3 + 1.$

What is the constant term of P(x)?

A)70　　　　　　　　B)72　　　　　　　C)78　　　　D)81

E)85

(Solution):

$$x^3 + 8 = 0 \Rightarrow x^3 = -8 \Rightarrow x = -2$$

$$P(0) = (-2)^6 - 2(-2)^3 + 1$$

=64+16+1

=81

(Answer D)

SUM &SUBSTRACTION ON POLYNOMIALS

(Example):

$P(x)=4x^3 + 6x - 1$

$Q(x) = 6x^3 + 2x + 9$

$\Rightarrow P(x) + Q(x) = ?$

(Solution):

$P(x)+Q(x)=4x^3 + 6x - 1 + 6x^3 + 2x + 9$

$=10x^3 + 5x + 8$

(Example):

$P(x)=4x^4 - 5x^3 - 7$

$Q(x)=-3x^4 + 6x^3 + 3$

$\Rightarrow P(x) - Q(x) = ?$

(Solution):

$P(x)-Q(x)= \left(4x^4 - 5x^3 - 7\right) - \left(3x^4 + 6x^3 + 3\right)$

$=4x^4 - 5x^3 - 7 + 3x^4 - 6x^3 - 3$

$=7x^4 - 11x^3 - 10$

(Example):

$P(x)=2x^3 - 4x^2+5x-1$

$Q(x)=6x^2 - 4x + 3$

1.P(x)+Q(x)=?

2.P(x)-Q(x)=?

(Solution):

1.P(x)+Q(x)= $\underset{P(x)}{\underline{2x^3 - 4x^2 + 5x - 1}} + \underset{Q(x)}{\underline{6x^2 - 4x + 3}}$

$=2x^3 + 2x^2 + x + 2$

2.P(x)-Q(x)= $\underset{P(x)}{\underline{2x^3 - 4x^2 + 5x - 1}} - \underset{Q(x)}{\underline{(6x^2 - 4x + 3)}}$

$=2x^3 - 4x^2 + 5x - 1 - 6x^2 + 4x - 3$

$=2x^3 - 10x^2 + 9x - 4$

(Example):

$P(x)=x^4 - x^3 + 2x^2 + 3x + 2$

$Q(x)=x^4 + 3x^2 - x + 5$

1.P(x)+Q(x)=?

2.P(x)-Q(x)=?

(Solution):

1.P(x)+Q(x)= $\dfrac{\overset{x^4 - x^3 + 3x + 2}{P(x)}}{} - \dfrac{\overset{x^4 + 3x^2 - x + 5}{Q(x)}}{}$

$= x^3 + 5x^2 + 2x + 7$

2.P(x)-Q(x)= $\dfrac{\overset{x^4 - x^3 + 2x^2 + 3x + 2}{P(x)}}{} - \dfrac{\overset{\left(-x^4 + 3x^2 - x + 5\right)}{Q(x)}}{}$

$= x^4 - x^3 + 2x^2 + 3x + 2 + x^4 - 3x^2 + x - 5$

MULTIPLICATION ON POLYNOMIALS

(Example):

$P(x) = x^2 - 3x + 4$

$Q(x) = x^3 - 2x^2 - 1$

$\Rightarrow P(x).Q(x) = ?$

(Solution):

$P(x).Q(x) = (x^2 - 3x + 4).(x^3 - 2x^2 - 1)$

$= x^5 - 2x^4 - x^2 - 3x^4 + 6x^3 + 3x + 4x^3 - 8x^2 - 4$

$= x^5 - 5x^4 + 10x^3 - 9x^2 + 3x - 4$

(Example):

$P(x) = x^3 - x^2, Q(x) = x^2 + ax - 9$

$P(x).Q(x) = x^5 + 4x^4 - 14x^3 + 9x^2 \Rightarrow a = ?$

$P(x).Q(x) = (x^3 - x^2).(x^2 + ax - 9)$

$= x^5 + ax^4 - 9x^3 - x^4 - ax^3 + 9x^2$

$= x^5 + (a - 1)x^4 - (9 + a)x^3 + 9x^2$

a-1=4$\Rightarrow a = 5$

(Example):

$P(x).Q(x) = x^4 + x^3 - 3x^2 - 4x - 4$

$P(x) = x^2 - 4, Q(x) = x^2 + ax + 1 \Rightarrow a = ?$

(Solution):

$P(x).Q(x) = (x^2 - 4).(x^2 + ax + 1)$

$= x^4 + ax^3 + x^2 - 4x^2 - 4ax - 4$

$= x^4 + ax^3 - 3x^2 - 4ax - 4$

$=_a x^4 + ax^3 - 3x^2 - 4ax - 4$

$=_a x^3 = x^3$

a=1

(Example):

$P(x) = 2x^3 - 4x^2 - 3x + 5$

$Q(x) = -3x^4 - 2$

$\Rightarrow P(x).Q(x) = ?$

(Solution):

$P(x).Q(x) = 2(x^3 - 4x^2 - 3x + 5).(-3x^4 - 2)$

13

$= -6x^7 - 4x^3 + 12x^6 + 8x^2 + 9x^5 + 6x - 15x^4 - 10$

$= -6x^7 + 12x^6 + 9x^5 - 15x^4 - 4x^3 + 8x^2 + 6x - 10$

DIVISION ON POLYNOMIALS

1.(Identity of Division):

$$\frac{P(x)}{Q(x)} \div \frac{Q(x)}{T(x)}$$

P(x)=Q(x).T(x)+K(x)

K(x) is the remainder

(Example):

$$\frac{P(x)}{4} \div \frac{x-3}{Q(x)}, \frac{Q(x)}{2} \div \frac{x+3}{T(x)} \Rightarrow \frac{P(x)}{?} \div x^2 - 9$$

A)3x+4   B)2x-2   c)2x+7

D)4x+1   E)5x+4

(Solution):

$$\frac{P(x)}{4} \div \frac{x-3}{Q(x)}$$

P(x)=(x-3).Q(x)+4

$$\frac{Q(x)}{2} \div \frac{x+3}{T(x)}$$

Q(x)=(x+3).T(x)+2

P(x)=(x-3)$\lfloor(x+3).T(x)+2\rfloor+4$

$=(x^2-9).T(x)+2x-6+4$

$=(x^2-9).$T(x)+2x-2

2x-2(Remainder)

Answer

b

RULE 1:

To find the remainder of P(x) divided by (x+a)is equal

to p(-a).Since (x+a) is a first order polynomial, the remainder must always be equal to real number.

P(x)=(x+a).T(x)+k

P(-a)=(-a+a).T(x)+K=K

$\Rightarrow P(-a)=K$

(Example):

P(x)=x^3-2x^2+ax+8

$\dfrac{P(x)}{18}\div x-2\Rightarrow a=?$

A)1   B)2   C)3

D)4   E)5

(Solution):

x-2=0$\Rightarrow x=2$

15

$P(2)2^3 - 2.2^2 + a.2 + 8$

18=8-8+2a+8

2a=10$\Rightarrow a = 5$

Answer B

(Example):

$P(3x+4)=x^3 + x^2 - x + 9 \Rightarrow$

$$\frac{P(x)}{?} \div x + 2$$

A)13   B)9   C)11

D)10   E)7

(Solution):

1.x+2=0$\Rightarrow x = -2$

2.3x+4=-2

3x=-6$\Rightarrow x = -2$

$P(3(-2)+4)=(-2)^3 + (-2)^2 - 2 + 9$

P(-2)=-8+4+2+9

=7

Answer C

(Example):

$P(4x-1)=x^3 - x^2 + 2x - 5 \Rightarrow$

$P(x)$

$$\frac{-}{?} \div \quad_{x-3}$$

A)2   B)1   C)0

D)-1   E)-3

(Solution):

1.x-3=0$\Rightarrow x = 3$

2.4x-1=3

4x=4$\Rightarrow x = 1$

$P(4.1-1)=1^3 - 1^2 + 2.1 - 5$

P(3)=1-1+2-5

=-3

                                        Answer E

(RULE2):(x-a).P(x)=Q(x)$\Rightarrow Q(a) = 0$

(Example):

(x-1)P(x)=$x^4 + ax^3 + 3x - 7 \Rightarrow a = ?$

A)2   B)3   C)5

D)7   E)8

17

(Solution):

x-1=0$\Rightarrow x = 1$

$1^4 + a.1^3 + 3.1 - 7 = (1 - 1)P(1)$

1+a+3-7=0

a=3

Answer B

(Example):

(x-2)P(x)=$x^4 - ax^2 + 2x + 8\Rightarrow$

$$\frac{P(x)}{?} \div x - 1$$

A)-4   B)-3   C)-1

D)3   E)5

(Solution):

1.x-2=0$\Rightarrow x = 2$

(2-2)P(2)=16-4a+4+8

0=28-4a

a=7

2.x-1=0$\Rightarrow x = 1$

(1-2)P(1)=1-7+2+8

-1P(1)=4

P(1)=-4

(Remainder)=P(1)=-4

Answer A

(Example):

$$9x^2 + 3x + 7 = (3x + 1).Q(x) + a \Rightarrow a = ?$$

A)1   B)2   C)4

D)7   E)9

(Solution):

$$3x+1=0 \Rightarrow x = -\frac{1}{3}$$

$$9\left(-\frac{1}{3}\right)^2 + 3\left(-\frac{1}{3}\right) + 7 = \left(3\left(-\frac{1}{3}\right) + 1\right).Q\left(-\frac{1}{3}\right) + a$$

$$9.\frac{1}{9} - 3.\frac{1}{3} + 7 = a$$

a=1-1+7

a=7

RULE 3

To Find the remainder of P(x) divided by $(x^n \pm a)$.

Insert$(x^n = \pm a)$ in the polynomial $P(x)$.

(Example):

$P(x) = x^{16} - 2x^{11} + 6x^6 + 3 \Rightarrow$

$\dfrac{P(x)}{?} \div x^5 + 2$

A)-28x+3   B)4x+9   C)-17x+21

D)-14x+7   E)21x+18

(Solution):

$x^5 + 2 = 0 \Rightarrow x^5 = -2$

$P(x) = x.x^{15} - 2x.x^{10} + 6x.x^5 + 3$

$= x\left(x^{5^2}\right) - 2x\left(x^{5^2}\right) + 6x\left(x^6\right) + 3$

(Remainder)$= x(-2)^3 - 2x(-2)^2 + 6x(-2) + 3$

$= -8x - 8x - 12x + 3$

$= -28x+3$

Answer A

(Example):

$$(x^2 + 4)P(x) + 8x = ax^2 + 2ax + b + 3 \Rightarrow b = ?$$

A)6   B)8   C)10

D)13   E)15

(Solution):

$$x^2 + 4 = 0 \Rightarrow x^2 = -4$$

((-4)+4)P(x)+8x=a(-4)+2ax+b+3

8x=2ax+b-4a+3

$2a=8 \Rightarrow a = 4$

b-4a+3=0

b-16+3=0

b=13

        Anwer D

(Example):

P(-1)=10

P(1)=4$\Rightarrow$

$$\frac{P(x)}{?} \div \frac{x^2 + 1}{x + 2}$$

A)2x+1   B)-5x+3   C)-3x+7

D)8x+1   E)-2x-4

(Solution):

$P(x)=(x^2+1)(x+2)+ax+b$

$X=1 \Rightarrow P(1)6+a+b=4 \qquad \Rightarrow a+b=-2$

$\Rightarrow P(-1)=2-a+b=10 \Rightarrow \dfrac{\begin{array}{c}b-a=8 \\ \hline 2b=6\end{array}}{}$

X=-1 $b=3$

$a=5 \Rightarrow ax+b=-5x+3$

Answer B

(Example):

$P(x)=x^3-3x^2+4x-9$

$\dfrac{\overset{-}{P(x)}}{?} \div x^2-x+1$

A)2x+5   B)x+9   C)x-7

D)3x-5   E)5x+8

(Solution):

$x^2-x+1=0 \Rightarrow x^2=x-1$

(if we write x-1 instead of x^2 in $P(x)$.)

$P(x)=x.x^2 - 3x^2 + 4x - 9$

(Remainder)=x(x-1)-3(x-1)+4x-9

$$=x^2 - x - 3x + 3 + 4x - 9$$

=x-1-6=x-7

Answer C

RULE 4

1.The remainder of P(x) and Q(x) divided by (x-a) are

Equal to A and B, respectively.

a)The remainder P(x) $\pm$ Q(x) devided by (x-a) is equal to

A $\pm$ B.

b)The remainder P(x).Q(x) divided by (x-a) equal to

A.B

2.The remainder of P(x) divided by (x-a) and (x-b) are

Equal to A and B, respectively. Then, the remainder

P(x) divided by (x-a).(x-b) is in the form of to mx+n).

(Example)

$$\frac{\overline{P(x)}}{3} \div x - 4$$

$$\frac{\overline{Q(x)}}{4} \div x - 4$$

$$\frac{x.P(x) + (x+1).Q(x).x^2 + 2x}{?} \overset{-}{} \div x - 4$$

A)56   B)48   C)46

D)44   E)40

(Solution):

1.x-4=0$\longrightarrow x = 4$

P(4)=3.Q(4)=4

2.(Remainder)=4P(4)+(4+1).Q(4)+$4^2 + 2.4$

=4.3+5.4+16+8

12+20+24

=56

Answer A

(Example):

$$\frac{P(x)}{12x+7} \overset{-}{} \div (3x-4)^2 \Rightarrow \frac{P(x)}{?} \overset{-}{} \div 3x - 4$$

A)15   B)17   C)19

D)21   E)23

(Solution):

$$\underset{3x-4=0}{\Rightarrow x = \frac{4}{3}}$$

P(x)=$(3x-4)^2 Q(x) + 12x + 7$

$$X=\dfrac{4}{3}$$

$$P(\dfrac{4}{3}) = \dfrac{(\dfrac{4}{3}\cdot3 - 4)^2}{0} - Q\left(\dfrac{4}{3}\right) + 12\dfrac{4}{3} + 7$$

(Remainder)=0+4.4+7

$$=23$$

Answer E

(Example):

$$\dfrac{\overset{-}{P(x)}}{5} \div x - 2$$

$$\dfrac{\overset{-}{Q(x)}}{9} \div x - 3 \Rightarrow \dfrac{\overset{-}{P(x)}}{?} \div (x - 2)(x - 3)$$

A)2x-7   B)4x+8   C)4x-3

D)8x+9   E)12x-1

(Solution):

K(x)=mx+n

$$\text{x-2=0} \Rightarrow x = 2 \Rightarrow 2m + n = 5$$

$$\begin{array}{c} \Rightarrow x = 3 \Rightarrow \\ \text{x-3=0} \end{array} \dfrac{- 3m + n = 9}{- m =- 4}$$

m=4(and)n=-3

K(x)=4x-3

(Example):

$P(x) = x^3 - x^2 + ax + b$

$$\frac{\overset{-}{P(x)}}{0} \div x^2 - 3x + 2 \Rightarrow a + b = ?$$

A)5   B)4   C)2

D)0   E)-2

(Solution):

$x^2 - 3x + 2 = (x-1)(x-2)$

1.X-1=0$\Rightarrow x = 1$

P(1)=1-1+a+b=0$\Rightarrow a = -b$

2.x-2=0$\Rightarrow x = 2$

P(2)=8-4+2a+b=0

2a+b+4=0

-2b+b+4=0

     b=4,a=-4

a+b=4(-4)=0

TEST WITH SOLUTION

1. $P(x)=x+4, Q(x)=x^2-5x \Rightarrow P(x)+Q(x)=?$

A) x^2-4x 　　　　　　 B) x^2-4 　　　　　　 C) x^2+4x+4

D) $(x-2)^2$ 　　　　　　　　　　　　 E) $(x+4)^2$

(Solution):

$$P(x)+Q(x)= \overset{x+4}{P(x)} + \overset{x^2-5x}{Q(x)}$$

$$=x^2-4x+4$$

$$=(x-2)^2$$

Answer D

2. $(x^3-4x^2+3x).(x^2-5x+1) = ... + a.x^4 + ...$

$\Rightarrow a=?$

A)-11 　　　　　　 B)-10 　　　　　　 C)-9

D)-8 　　　　　　 E)-7

(Solution):

$(x^3-4x^2+3x).(x^2-5x+1)$

$-5x^4-4x^4 = ax^4$

$-9x^4 = ax^4$

a=-9

Answer C

3.P(x-4)=$2x^2 + 3x + 4 \Rightarrow P(1) = ?$

A)69   B)70   C)71

D)72   E)73

(Solution):

P(x-4)=$2x^2 + 3x + 4$

$\Rightarrow x - 4 = 1 \Rightarrow x = 5$

P(5-4)=$2.5^2 + 3.5 + 4$

=2.25+15+4

=50+15+4

=69

Answer A

4.P(x-3)=$x^3 + 2x^2 - x + a, \quad P(-1) = 5 \Rightarrow a = ?$

A)-10   B)-9   C)-8

D)-7   E)-6

(Solution):

P(x-3)=$x^3 + 2x^2 - x + a$

28

$X=2 \Rightarrow$

$P(2-3)=2^3 + 2.2^2 - 2 + a$

$P(-1)=8+2.4-2+a$

$P(-1)=8+8-2+a$

$P(-1)=14+a$

$5=14+a \Rightarrow a = -9$

Answer B

5. $P(x^2) = x^4 - 1. Q(\sqrt{x}) = x + 1 \Rightarrow P(x).Q(x) = ?$

A) $x^4 + 1$ B) $x^4 - 1$ C) $x^2 - 1$

D) $x^2 + 1$ E) $x^5 - 1$

(Solution):

$P(x^2) = x^{2^2} - 1$ $Q(\sqrt{x^2}) = x^2 + 1$

$P(x) = x^2 - 1$ $Q(x) = x^2 + 1$

$P(x).Q(x) = (x^2 - 1).(x^2 + 1)$

$= x^4 + x^2 - x^2 - 1$

$= x^4 - 1$

Answer B

6. $m \in Z$

$P(x)=x^{2m+2}+x^{2m+1}+2 \Rightarrow P(-1) = ?$

A)2   B)3   C)4

D)5   E)6

(Solution):

$P(x)=x^{2m+2}+x^{2m+1}+2$

$P(-1)=(-1)^{2m+2}+(-1)^{2m+1}+2$

=1+(-1)+2

=2

Answer A

7.$P(2-3x)=-2x^7+5x^3+2x^2+8 \Rightarrow P(5) = ?$

A)3   B)4   C)5

D)6   E)7

(Solution):

$P(2-3x)=-2x^7+5x^3+2x^2+8$

$X=-1 \Rightarrow$

$P(2-3(-1))=-2.(-1)^7+5.(-1)^3+2.(-1)^2+8$

P(2+3)=-2.(-1)+5.(-1)+2.1+8

P(5)=2-5+2+8

=7

Answer E

8.$P(x)=2x^4 - ax^3 + x^2 - (3+b)x + 1$

$Q(x)=(c+1)x^4 - 2x^2 + 2x + 3,$

$P(x)+Q(x)= -x^2 + 4 \Rightarrow$ $a+b+c = ?$

A)-7   B)-6   C)-5

D)-4   E)-3

(Solution):

$P(x)+Q(x)=(c+1+2)x^4 - ax^3 - x^2 + (2-3-b)x + 4$

$=(c+3)x^4 - ax^3 - x^2 + (-1-b)x + 4$

$(c+3)x^4 - ax^3 - x^2 + (-1-b)x + 4 = x^2 + 4$

C+3=0   -a=0   -1-b=0

c=-3   a=0   b=-1

a+b+c=0+(-1)+(-3)

=-4

Answer D

9.$P(x,y)=2x^2y^2 - 3xy^2 - 6x + 1 \Rightarrow p(3,\sqrt{2}) = ?$

A)1   B)2   C)3

D)4   E)5

31

(Solution):

$$P(3.\sqrt{2}) = 2.3^2.(\sqrt{2})^2 - 3.3(\sqrt{2})^2 - 6.3 + 1$$

=2.9.2-9.2-18+1

=36-18-18+1

=1

Answer A

10.P(x)=$x^3 + ax^2 - bx + 1$,$P(1) = 10 \Rightarrow a - b = ?$

A)4   B)5   C)6

D)7   E)8

(Solution):

$$P(1)=1^3 + a.1^2 - b.1 + 1 = 10$$

1+a-b+1=10

a-b=8

Answer E

11.P(x)= $x^4 - 5x^2 + 4 \Rightarrow \dfrac{P(x) - Q(x)}{(x^2 + 1).(x + 1)}$

Q(x)=$x^3 - 4x^2 + x + 6$

A)x-1   B)x+1   C)x-2

D)x+3   E)x-3

(Solution):

$$\frac{P(x) - Q(x)}{(x^2 + 1).(x + 1)}$$

$$= \frac{x^4 - 5x^2 + 4 - (x^3 - 4x^2 + x + 6)}{(x^2 + 1).(x + 1)}$$

$$= \frac{x^4 - 5x^2 + 4 - x^3 + 4x^2 - x - 6}{(x^2 + 1).(x + 1)}$$

$$= \frac{x^4 - x^2 - 2 - x^3 - x}{(x^2 + 1).(x + 1)}$$

$$= \frac{(x^2 - 2).(x^2 + 1) - x(x^2 + 1)}{(x^2 + 1).(x + 1)}$$

$$= \frac{x^2 - x - 2}{x + 1}$$

$$= \frac{(x - 2)(x + 1)}{x + 2}$$

=x-2

Answer E

12.P(x)=(2k-1)x^3 + $(k + 1)x^2 - x + k$

P(-2)=36$\Rightarrow k = ?$

A)-3   B)-2   C)0

D)2   E)3

(Solution):

$P(-2)=(2k-1).(-2)^3 + (k+1)(-2)^2 - 2 + k$

=(2k-1).(-8)+(k+1).4+2+k

=-16k+8+4k+4+2+k

=-11k+14

-11k+14=36

-11k=22

K=-2

Answer B

13. $(x^2 + x + 1).P(x + 5) = x^3 - 1 \Rightarrow P(x) = ?$

A)x+6   B)x-5   C)x-6

D)x+5   E)x-1

(Solution):

$(x^2 + x + 1).P(x + 5) = x^3 - 1$

$P(x+5) = \dfrac{x^3 - 1}{x^2 + x + 1}$

$P(x+5) = \dfrac{(x-1).(x^2 + x + 1)}{x^2 + x + 1}$

P(x+5)=x-1

P(x-5+5)=x-5-1

P(x)=x-6

Answer C

14. $m, n \in Z^+$

$$4x^2 - mx + 4 = (2x - n)^2 \Rightarrow m + n = ?$$

A)10   B)12   C)14

D)16   E)18

(Solution):

$$4x^2 - mx + 4 = (2x - n)^2$$

$$4x^2 - mx + 4 = 4x^2 - 4nx + n^2$$

$-mx+4=-4nx+n^2$

$-m=-4n$ (and) $n^2 = 4, n = 2$

$n=2 \Rightarrow m = 8 \Rightarrow m + n = 10$

Answer A

15. $x^3 + ax^2 + bx + c = (x - 2).(x + 4).(x + 1)$

$\Rightarrow a.b.c = ?$

A)96   B)108   C)120

D)132   E)144

(Solution):

$$x^3 + ax^2 + bx + c = (x - 2).(x + 4).(x + 1)$$
$$= (x^2 + 4x - 2x - 8).(x + 1)$$
$$= (x^2 + 2x - 8)(x + 1)$$
$$= x^3 + x^2 + 2x^2 + 2x - 8x - 8$$
$$= x^3 + 3x^2 - 6x - 8$$
$$x^3 + ax^2 + bx + c = x^3 + 3x^2 - 6x - 8$$
$$a=3, \quad b=-6, c=-8$$

a.b.c=3.(-6).(-8)

=144

Answer E

16.P(x)=-3.x^{40} + 12.x^{20} - 12 $\Rightarrow P(\sqrt[5]{2}) = ?$

A)-768   B)-640   C)-612

D)-588   E)-542

(Solution):

P(x)=3.x^{5^6} + 12.$\left(x^{5^4}\right)$ - 12

X=$\sqrt[5]{2}$ $\Rightarrow$

$P(\sqrt[5]{2}) = -3((\sqrt[5]{2^5})^8 + 12((\sqrt[5]{2^5})^2 - 12$

$= -3.2^8 + 12.2^4 - 12$

$= -588$

Answer D

17. $P(2x+5) = (2x^2 + 3x - 1).Q(x + 1)$

$Q(-1) = 3 \Rightarrow P(1) = ?$

A)1   B)2   C)3

D)4   E)5

(Solution):

$P(2x+5) = (2x+3x-1).Q(x+1)$

$X = -2$ (for)

$P(2.(-2)+5) = (2.(-2)^2 + 3.(-2) - 1).Q(-2+1)$

$P(1) = (2.4-6-1).Q(-1)$

$P(1) = Q(-1)$

$P(1) = 3$

Answer C

18. $P(x) = ax^2 + bx + c, P(2) = 0, P(3) = 0$

$\Rightarrow \dfrac{a}{b} = ?$

A) $-\dfrac{1}{10}$ B) $-\dfrac{1}{5}$ C) $\dfrac{1}{5}$

D) $\dfrac{1}{10}$ E) $\dfrac{1}{2}$

(Solution):

$P(x)=ax^2 + bx + c$, $P(2) = a.2^2 + b.2 + c$

$=4a+2b+c \Rightarrow 4a + 2b + c = 0$

$P(3)=a.3^2 + b.3 + c$

$=9a+3b+c \Rightarrow 9a + 3b + c = 0$

$$4a + 3b + c = 0$$

....

$$- 9a - 3b - c = 0$$

$$4a + 2b + c = 0$$

....

$$- 5a - b = 0$$

$$-5a = b$$

$$\dfrac{a}{b} = -\dfrac{1}{5}$$

Answer B

19.$P(x)= - x^2 + 3x.Q(x) = 5x^2 - x + 2$

$\Rightarrow P[Q(1)] + Q[P(-1)] = ?$

A)36   B)48   C)50

D)56   E)68

(Solution):

$Q(1)=5.1-1+2$ $,P(-1)=-(-1)^2+3.(-1)$

$Q(1)=6$ $,P(-1)=-1-3$

                   $P(-1)=-4$

$P[Q(1)]+Q[P(-1)]=P(6)+Q(-4)$

$=-6^2+3.6+5.(-4)^2—4+2$

=-36+18+5.16+4+2

=-36+18+80+4+2

=68

Answer E

20. $P(x+3)=2x^3+ax^2+x+1, P(4)=10 \Rightarrow a=?$

A)-6   B)-2   C)2

D)6   E)10

(Solution):

$P(x+3)=2x^3+ax^2+x+1, x=1(for)$

$P(1+3)=2.1^3+a.1^2+1+1$

P(4)=2+a+2

P(4)=4+a

4+a=10

a=6

Answer D

21.$P(x^2 + 2) = x^8 + x^6 + ax^4 + 3$

$P(0)=47 \Rightarrow a = ?$

A)7   B)8   C)9

D)10   E)11

(Solution):

P(0)=47

$x^2 + 2 = 0 \Rightarrow x^2 = -2$

$P(-2+2)=(-2)^4 + (-2)^3 + a(-2)^2 + 3$

47=16-8+4a+3

47=11+4a

36=4a

a=9

Answer C

22.P(x+1)-5=xP(x)+Q(x)

P(1)=35

$\Rightarrow Q(0) = ?$

A)30   B)37   C)42

D)47   E)55

(Solution):

P(1)=35

Q(0)=?$\Rightarrow$

P(1)-5=0.P(0)+Q(0)

35-5=Q(0)

Q(0)=30

Answer A

QUESTIONS

1.P(x)=$x^3 + 3x^2 + x, Q(x) = 5x^2 + bx + 1$

P(x)+Q(x)=$x^3 + 8x^2 + 5x + 1 \Rightarrow b = ?$

A)1   B)2   C)3

D)4   E)5

(Solution):

$P(x)+Q(x)=x^3 + 3x^2 + x + 5x^2 + bx + 1$

$=x^3 + 8x^2 + bx + x + 1$

$= x^3 + 8x^2 + (b+1)x + 1$

$x^3 + 8x^2 + 5x + 1 = x^3 + 8x^2 + (b+1)x + 1$

$5=b+1 \Rightarrow b = 4$

Answer D

2.$P(x)=x+1, Q(x)=x^2 + x$

$\Rightarrow P(x) + Q(x) = ?$

A)$x^2 - 2x$ B) $x^2 + 2x$ C)(x-1)(x+1)

D)$(x+1)^2$ E)$(x-1)^2$

(Solution):

$P(x)+Q(x)=x+1+x^2 + x$

$\qquad = x^2 + 2x + 1$

$\qquad = (x+1)^2$

Answer D

3.$P(x)=x^2 + 5x - 3, Q(x) = x + 1$

$\Rightarrow P[Q(-1)] + Q[P(2)] = ?$

A)14   B)11   C)9

D)-6   E)-4

(Solution):

$Q(-1)=-1+1=0, P(2)=2^2 + 5.2 - 3$

$P(2)=4+10-3$

$P(2)=11$

$P[Q(-1)] + Q[P(2)] = P(0) + Q(11)$

$$=0^2 + 5.0 - 3 + 11 + 1$$

$$=-3+12=9$$

Answer C

4. $P(x)=ax^2 + bx + c$, $P(1) = 0, P(2) = 0$

$$\Rightarrow \frac{b}{a} = ?$$

A)3   B)2   C)0

D)-2   E)-3

(Solution):

$P(1)=a.1^2 + b.1 + c = 0 \Rightarrow -1/a + b + c = 0$

$$2^2 + b.2 + c = 0 \Rightarrow$$

$$\begin{array}{r} 4a + 2b + c = 0 \\ \hline -a - b - c = 0 \\ + 4a + 2b + c = 0 \\ \hline 3a + b = 0 \end{array}$$

P(2)=a

$$b = -3a$$

$$\frac{b}{a} = -3$$

Answer E

5. $x^3 + ax^2 + bx + c = (x + 2).(x - 3).(x - 1)$

$$\Rightarrow \frac{c}{a} = ?$$

A)3   B)2   C)1

D)-1   E)-3

(Solution):

$$x^3 + ax^2 + bx + c = (x + 2)(x - 3)(x - 1)$$

$$= (x^2 - 3x + 2x - 6)(x - 1)$$

$$= x^3 - x^2 - x^2 + x - 6x + 6$$

$$x^3 + ax^2 + bx + c = x^3 - 2x^2 - 5x + 6$$

a=-2,b=-5,c=6, $\dfrac{c}{a} = \dfrac{6}{-2} = -3$

Answer E

6.a,b $\in R$ $4x^2 - 5x + b = (2x - a)^2 = 0$

a+b=?

A) $\dfrac{5}{4}$

B) $\dfrac{25}{8}$

C) $\dfrac{25}{16}$

D) $\dfrac{35}{8}$

E) $\dfrac{45}{16}$

(Solution):

$4x^2 - 5x + b = (2x - a)^2$

$4x^2 - 5x + b = 4x^2 - 4ax + a^2$

-5=4a

$a = \dfrac{5}{4}$

$b = a^2$

$b = \left(\dfrac{5}{4}\right)^2 \Rightarrow b = \dfrac{25}{16}$

$a+b = \dfrac{5}{4} + \dfrac{25}{16} = \dfrac{20}{16} + \dfrac{25}{16} = \dfrac{45}{16}$

(4)

Answer E

7.
$$\dfrac{\overset{3x^2 - 2mx^2 - nx - 2}{-}}{0} \div \dfrac{x^2 - x - 2}{Q(x)} \Rightarrow m = ?$$

(Solution):

$3x^2 - 2mx^2 - nx - 2 = (x - 2)(x + 1).Q(x)$

X=2 $\Rightarrow 24 - 8m - 2n - 2 = 0$

$\Rightarrow 11 = 4m + n$

X=-1 $\Rightarrow -3 - 2m + n - 2 = 0$

$\Rightarrow 2m - n = -5$

4m+n=11

2m-n=-5

....

6m=6

m=1

Answer D

1. $(2x + 1)^3 = 8x^3 + ax^2 + bx + c \Rightarrow a + b + c = ?$

A)12   B)14   C)16

D)19   E)21

2. $(a+x).(x^2 + ax + 2b) = x^3 + 3x^2 + 5x + c + 2$

$\Rightarrow 2a + c = ?$

A)5   B)6   C)7

D)8   E)9

3. $P(x) = 3x^5 - x - 1 \Rightarrow P(1) = ?$

A)0   B)1   C)2

D)3   E)4

4. $P(1-x) = 2x^2 + 2 \Rightarrow P(X) = ?$

A) $2x^2 - 4$ B) $2x^2 + 4x$ C) $2x^2 - 4x$

D) $2x^2 + 4x + 4$ E) $2x^2 - 4x + 4$

5. $P(x-2) = ax+b-2a \Rightarrow P(2 - x) = ?$

A)-ax-b   B)ax-b   C)-ax+b+2a

D)-ax+2b+a   E)ax-b+2a

6. $P(x-2) = ax^6 - bx^3 + cx^2 - dx - 9$

$P(-3) = 6 \Rightarrow a + b + c + d = ?$

A)15   B)16   C)17   D)18   E)19

7. $P(x^3 + 1) = x^8 ++ a.x^6 - 3x^4 + 2$

$P(0) = 20 \Rightarrow a = ?$

A)20   B)15   C)10

D)5   E)2

8. $P(x) = x^7 + 4x + a, P(2) = 0 \Rightarrow ?$

A)-136   B)-120   C)-90

D)136   E)140

9. $P(x+2) = 2x^4 - 3x - 2$

$Q(x) = 3x^2 - 2x + 2 \Rightarrow P(0).Q(0) = ?$

A)56   B)64   C)72

D)76   E)80

10. $P(x) = (3x-4).Q(x)+3,$ $P(4) = 19 \Rightarrow Q(4) = ?$

A)1   B)2   C)3

D)4   E)5

11.P(x-2)=(x-2).Q(x+2)+x+3,P(2)=15

$\Rightarrow Q(6) = ?$

A)1   B)2   C)3

D)4   E)5

12.P(x)=ax+b$\Rightarrow P(1) - P(2) = ?$

A)-a   B)-b   C)2b

D)a   E)b

13.(x+3).P(x+3)+2=$x^3 ax + 5 \Rightarrow P(3) = ?$

A)-2   B)-1   C)0   D)1   E)2

14.P(x+1)=$(x^2 + 2x - 1).Q(x) + x - 2$

P(2)=11$\Rightarrow Q(1) = ?$

A)2   B)3   C)4

D)5   E)6

15.P(x+3)=$x^3 - 2x - 3 \Rightarrow P(-2) = ?$

A)-156   B)-144   C)-118

D)118   E)144

16. $P(x.y)=x^3.y - x^2.y^2 + 2.y^4$

$\Rightarrow P(\sqrt{5}. - \sqrt{5}) = ?$

A)-25   B)-5   C)0

D)-2   E)25

17. $P(x^2) = x^4 + 5x^2 + 8 \Rightarrow P(-3) = ?$

A)1   B)2   C)3

D)-2   E)-3

18. P(2x+1)=2x+5 $\Rightarrow P(x) = ?$

A)x+2   B)x+4   C)x+5

D)2x+1   E)x.(x+3)

19. P(x)=$2x^2 - 2x + 1$

P(x+2)=P(x-2)$\Rightarrow x = ?$

A)$-\dfrac{1}{4}$ B) $-\dfrac{1}{8}$ C) $\dfrac{1}{8}$

D) $\dfrac{1}{4}$ E) $\dfrac{1}{2}$

20. P(x)=$x^3 + 6x^2 + 12x + 8$ $\Rightarrow P(x - 2) = ?$

A)$x^3 + 2$ B) $x^3 + 8$ C) x^3

D) $x^3 - 2$ E)$x^3 - 8$

50

21. $\dfrac{P(x+2)}{Q(X)} = 3x^2 - x - 15, \qquad Q(-3) = 4$

$\Rightarrow P(-1) = ?$

A)30   B)40   C)50   D)2

22. $P(x-1) = 2x^2 + ax + b$

$P(x+1) = 2x^2 + x + 1 \Rightarrow a.b = ?$

A)-49   B)-21   C)-15

D)2   E)1

23. $P(x) = (x-2)^3 \Rightarrow P(\sqrt[3]{3} + 2) = ?$

A)-3   B)-4   C)3   D)2   E)1

24. $P(x+1) = x^2 - 4x + 7 \Rightarrow P(1) = ?$

A)9   B)8   C)7   D)6   E)5

25. $P(x) = x^3 - 3x^2 + 3x - 1 \Rightarrow P(x+1) = ?$

A)x^3 B)$2x - x^3$ C)$x^3 - 1$ D)
$x^3 + 2x$ E)$1 - x^3$

(Answers)

1.D	2.B	3.B	4.E	5.C	6.A
7.A	8.A	9.C	10.B	11.D	12.A
13.D	14.E	15.C	16.A	17.B	18.B
19.E	20.C	21.D	22.A	23.C	24.C
25.A					

1.P(x)=$x^2 + 2x$

Q(x)=x-3$\Rightarrow P[Q(4)] + 2.Q[P(1)] = ?$

A)0   B)1   C)2

D)3   4)4

2.P(2x+1)=4x+5$\Rightarrow P(3) = ?$

A)-1   B)4   C)5

D)9   E)11

3.P(x+2)=$4x^2 + mx + 5$

P(4)=51$\Rightarrow m = ?$

A)2   B)4   C)5

D)6   E)7

4.P(x)=$2x^3 + (m - 2)x^2 + 4x + 5$

  P(x)=(x-3).Q(x)+116

A)4   B)5   C)6

D)7   E)8

5.P(x+1)=$x^3 - 4x^2 + x + m$

P(x+1)=(x-2).Q(x)+10$\Rightarrow m = ?$

A)6   B)10   C)12

D)16   E)18

6.$P(x+1)=x^2 - 3x \Rightarrow P(x) = ?$

A)$x^2 - 2x + 1$ B)$x^2 - 5x + 4$ C)2
$x^2 - 6x + 3$

D)$x^2 - 4x + 2$ E)$x^2 - 5x - 2$

7.$P(2x+4)=x^3 - 4x^2 + 1 \Rightarrow P(5) = ?$

A)1   B)$\dfrac{1}{2}$ C)$\dfrac{1}{4}$

D)$\dfrac{1}{4}$ E)$\dfrac{1}{64}$

8.$P(x)=2\sqrt{2}x + 12 \Rightarrow P(\sqrt{2}) = ?$

A)5   B)8   C)16

D)$4\sqrt{2}$ E)$\sqrt{2} + 12$

9.$ax^3 + 2x^2 + bx + c = (x + 1).(x - 2).(x + 3)$

$\Rightarrow a + b + c = ?$

A)-9   B)-10   C)13

D)14   E)21

10. $P(x,y)=4x^2y^3 - 2x^2 + y^2x + 16 \Rightarrow P(-1,1) = ?$

A)-2   B)7   C)17

D)19   E)20

11. $P(x)=x^3 - ax^2 + bx + 7$

$P(2)=0$

$P(1)=4 \Rightarrow a = ?$

A)$\dfrac{1}{2}$ B) $\dfrac{2}{3}$ C) $\dfrac{7}{2}$

D)4   E)16

12. $P(x)=9x^2 + 8x$

$Q(x)=4x+3 \Rightarrow P[Q(2)] - Q[P(3)] = ?$

A)211   B)105   C)754   D)801
E)902

13. $P(x)=6x^2 + 4x + 3 + b$

$P(x)=(x-1).Q(x)+27 \Rightarrow b = ?$

A)12   B)13   C)14

D)15   E)16

14. $P(x)=2x^2 + 4x - 10 \Rightarrow P(-3) = ?$

A)-5   B)13   C)14

D)15   E)16

15. $P(x)=4x^2 + 7x - 8$

$Q(x)=3x^3 + 5x^2 - 4x - 7$

$P(x)+Q(x)=T(x) \Rightarrow T(1) = ?$

A)-2   B)-1   C)0

D)15   E)24

16. $P(x-2)=4x^3 + 5x^2 - 6 \Rightarrow P(-1) = ?$

A)-4   B)3   C)2

D)4   E)12

17. $P(x)=2x^2 + ax + b$

$P(1)=0 \Rightarrow a + b = ?$

A)-2   B)4   C)10

D)12   E)14

18. $P(x+2)=3x^3 + ax^2 + x + 1$

$P(3)=8 \Rightarrow a = ?$

A)0   B)1   C)2   D)3   E)4

19. $P(x)= x^2 + 4x - 5 \Rightarrow \dfrac{P(x) + Q(x)}{(x-1)(x+1)} = ?$

Q(x)=4-4x

A)0   B)1   C)2

D)3   E)4

20.P(x)=4x+3

$Q(x)=2x+1 \Rightarrow P(x).Q(x) = ?$

A)$2x^2 + 5x + 4$ B) $8x^2 + 10x + 3$ C)

$5x^2 - 2x + 3$

D) $- 8x^2 - 5x + 3$ E) $2x^2 - 10x + 5$

21.P(x)=$x^2 + 4x \Rightarrow P(P(- 4)) = ?$

A)0   B)1   C)2   D)3

E)4

22.P(x)=$(x^2 - 3x + 2),$ $Q(x) = 4x^3 + x^2$

$\Rightarrow P(1) + Q(2) = ?$

A)32   B)36   C)47   D)54

E)60

23.P(x)= $x^3 + 1$ $\dfrac{\overset{P(x)}{-}}{2} \div \dfrac{x - 1}{Q(X)}$

$\Rightarrow Q(0) = ?$

A) $\dfrac{1}{2}$ 　　　　　 B)1 　　　 C)- $\dfrac{1}{2}$ 　　 D)-1 　　 E)- 2

24. P(x)=$ax^2 + bx + c$

P(0)=2 　　　　　　 $\Rightarrow a + b + c = ?$

P(1)=8

A)8 　　　　 B)6 　　　　 C)4

D)2 　　　 E)0

(Answers)					
1.D	2.D	3.E	4.D	5.D	6.E
7.D	8.C	9.A	10.C	11.C	12.C
13.C	14.B	15.C	16.B	17.A	18.D
19.B	20.B	21.A	22.B	23.D	24.C

1.$P(x+1)=x^2 - 4x + 5 \Rightarrow P(2 - x) = ?$

A)$x^2 - x + 2$ 　　　　　　　　　B) $x^2 + 2x + 2$

C) $x^2 + x - 2$

D) $x^2 - x + 1$ 　　　　　　　　　E) $x^2 - 2x - 1$

2.$P(x-2)=x^2 + x - 6 \Rightarrow P(x + 2) = ?$

A) $x^2 + 9x + 14$ 　　　　　　　B) $x^2 + 7x + 7$ 　　　　　　C)
$x^2 - 9x - 14$

D) $x^2 - 7x + 7$ 　　　　　　　　E) $x^2 + 8x$

3.$P(x,y)=x^3 + 3x^2y + 3xy^2 + y^3$

$\Rightarrow P(\sqrt[3]{4} + 2, \sqrt[3]{4} - 2) = ?$

A)64 　　　　　　B)32 　　　　　　C)16 　　　　　　D)8
E)4

4.$P(x,y)=x^2 - 4xy + 4y^2$

$\Rightarrow P(2\sqrt{2}, 2\sqrt{2}) = ?$

A)12 　　　　　　B)8 　　　　　　C)4 　　　　　　D)2
E)0

5.$P(x+2)=ax^2 + 6x^2 + 3x - 2$

$P(1)=2a-5 \quad \Rightarrow P(3) = ?$

A)1   B)-1   C)-2   D)5

E)9

6. $P(x+1)=mx^2 + x + 1$

$P(0)=3 \quad \Rightarrow P(-1) = ?$

A)11   B)10   C)7   D)5

E)1

7. $P(x+2)=-3x^4 + 2x^2 + 4x - 2$

$\Rightarrow P(1) - P(0) = ?$

A)43   B)18   C)8   D)4   E)2

8. $P(2x-3)=-8x^3 + 2x + 4 \quad \Rightarrow P(0) = ?$

A)-28   B)-20   C)-12   D)-4   E)8

9. $P(\dfrac{x}{2}) = 5x^5 + 4x^3 + 15x - 7$

$\Rightarrow P(0) = ?$

A)-5   B)-6   C)-7   D)-8   E)-9

10.$P(x-2)= \dfrac{x^2 - 5x - 8}{Q(2x - 1)} - x$

$\quad P(-1)=23 \quad \Rightarrow Q(1) = ?$

A)18   B)7   C)-3   D)$-\dfrac{1}{2}$ E) $-\dfrac{7}{2}$

11.$P(x-1).Q(x+3)=x^3 - 3x^2 - 4$

$\quad P(0)=-3 \Rightarrow \quad Q(4) = ?$

A)-3   B)-2   C)0   D)1   E)2

12.$P(x-2)=2x^2 - 3x - 6$

$\quad \Rightarrow P(\sqrt{2}) = ?$

A)$-\sqrt{2} + 1$ B)$-3\sqrt{2}$ C)$2\sqrt{2}$ D)$4\sqrt{2} + 1$
E)$5\sqrt{2}$

13.$P(x+1)=x^2 + x + 1$

$\quad P(\sqrt{3}) = ?$

A)$5\sqrt{3}$ B)$3+\sqrt{3}$ C)$2+\sqrt{3}$ D)$\sqrt{3}$
E)$4-\sqrt{3}$

14.$P(x-2)=x^2 - 4x + 4$

$\quad \Rightarrow P(x + 2) = ?$

A) $x^2 + 4x + 4$ B) $x^2 - 3x + 4$ C)

$x^2 - 2x - 3$

D) $x^2 + 4$ E) $x^2 - 5x + 3$

$$\frac{25x - 9}{x^2 + 4} = \frac{A}{x - 1} + \frac{B}{x + 1}$$

15.

$\Rightarrow A + B = ?$

A)5 B)8 C)12 D)17
E)25

16.P(x)= $x^2 - 3x + 1$ $\Rightarrow P(2x - 4) = ?$

A) $3x^2 - 8x - 7$ B) $4x^2 + 5x + 9$ C)

$4x^2 - 22x + 29$

D) $5x^2 + 4x + 10$ E) $2x^2 + x + 17$

17.P(3x+8)=$5x^2 + 3x - 4$

$\Rightarrow P(-1) = ?$

A)32 B)24 C)18 D)8
E)2

18.P(1-x)=-3x+5 $\Rightarrow P[P(2)]=?$

A)24 B)26 C)32 D)48
E)57

19.$P(2x+2)=x^2 + 1 \Rightarrow P(P(0)) = ?$

A)-3   B)1   C)5   D)8
E)17

$$\frac{-7x+7}{20.\,x^2-5x+6} = \frac{A}{x-3} + \frac{B}{X-2} \Rightarrow A+2B = ?$$

A)-5   B)0   C)7   D)17
E)20

$$\frac{P(x+1)}{21.\,Q(2x+1)} = x^2 + 10x + 3$$

$P(0)=-18 \Rightarrow Q(-1) = ?$

A)-1   B)3   C)4   D)5
E)6

22.$P(2x-3)=4x-6 \Rightarrow P(2x) = ?$

A)2x   B)4x   C)6x   D)8x
E)9x

23.$P(x-2)=Q(x).(x^2 - x - 2) + x^3$

$\Rightarrow P(0) = ?$

A)2   B)4   C)6   D)8
E)10

24. $\dfrac{\overline{P(x)}}{13} \div \dfrac{x+2}{x^2-3} \Rightarrow \dfrac{P(0)}{P(-2)} = ?$

A) $\dfrac{7}{13}$ 　　　　　 B) 3 　　　　　 C) $\dfrac{10}{3}$ 　　　　 D) $\dfrac{13}{2}$

E) 10

(Answers)					
1.B	2.A	3.B	4.B	5.E	6.A
7.A	8.B	9.C	10.D	11.E	12.E
13.E	14.A	15.E	16.C	17.A	18.B
19.B	20.B	21.B	22.B	23.D	24.A